Keep Your Android Phone Running All Day

STEVEN MANN

Keep Your Android Phone Running All Day

Introduction

One of the biggest complaints on all of these new fancy smart phones is that the battery gets drained very quickly. I noticed this from day one with my recent Android phone purchase. However, after a day of tweaking every single possible setting to conserve power without losing performance or functionality, I can now run my Android phone for over 24 hours before needing a charge. If you notice in the screenshot below, before I plugged my phone into the computer, I was running for almost 25 hours and still had 20-30% battery life remaining.

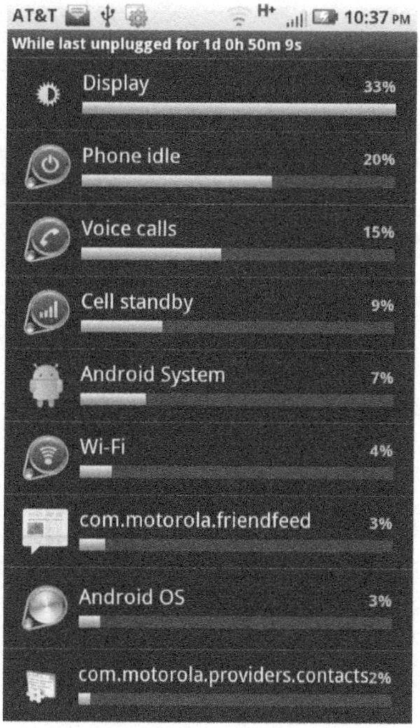

This was after a day of getting social networking updates from Twitter, FaceBook, and LinkedIn, reading and sending emails, texting, voice calls, browser usage, etc. So I was using it all day and still had a few hours if I so needed them before it would die.

I was able to tweak my phone and make a huge difference. Some settings are very obvious while other tricks, tips, and settings may not be. I thought that a guide book like this would help others and provide people with some things to think about or change in order to preserve battery life and enjoy their phone experience -without having to run for a charger.

Battery Manager

The obvious first place to start is the Battery Manager settings on your Android phone. Accessing the main settings of your phone using the device's settings button (on the device not on the screen) displays the Battery Manager in the list of settings:

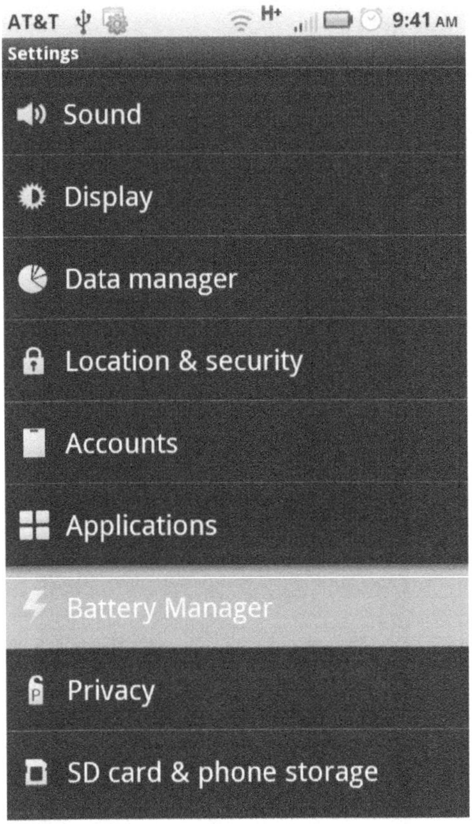

Selecting the Battery Manager option opens the Battery Manager screen and displays amount of battery life remaining both in a percentage as well as a visual picture. The first key option on this screen is the Battery Mode selection at the bottom.

Note: *Some phones may have the battery management option in the About Phone section.*

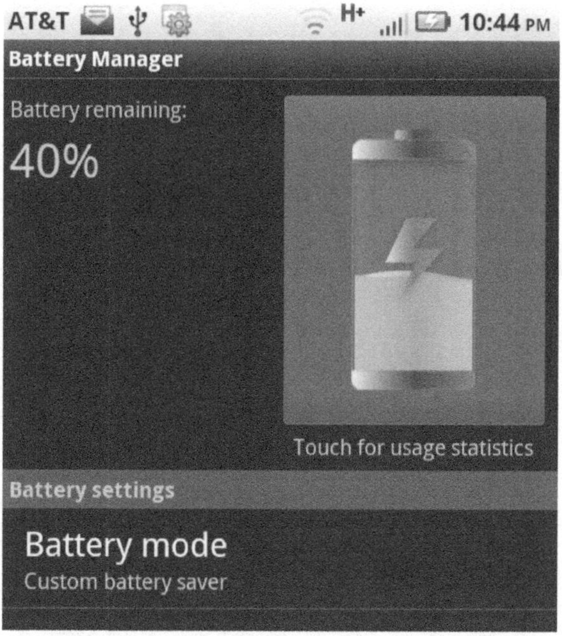

Select the Battery Mode to open the Battery Mode selection screen:

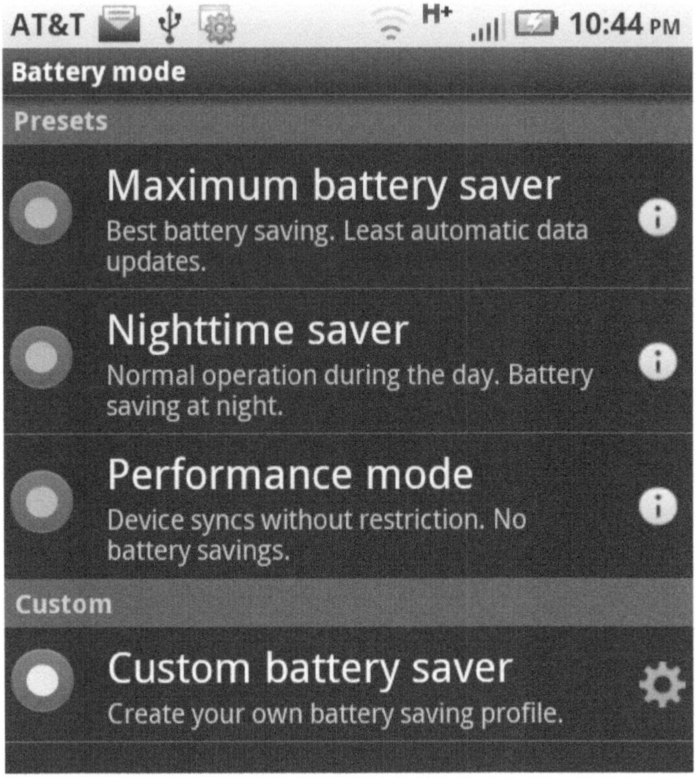

The Maximum preset seems a little too harsh and the Performance mode seems like your battery will run out in a few hours. Therefore I first liked the Nighttime saver. Since I use my phone as my morning alarm, I need it to be on overnight. Therefore since I didn't really need it for anything else, the Nighttime saver option was probably good.

This option may be good for you too but I decided to use the Custom batter saver option. To do this, select the little gear icon to the right of the Custom batter saver option - the Custom Batter Saver profile screen appears:

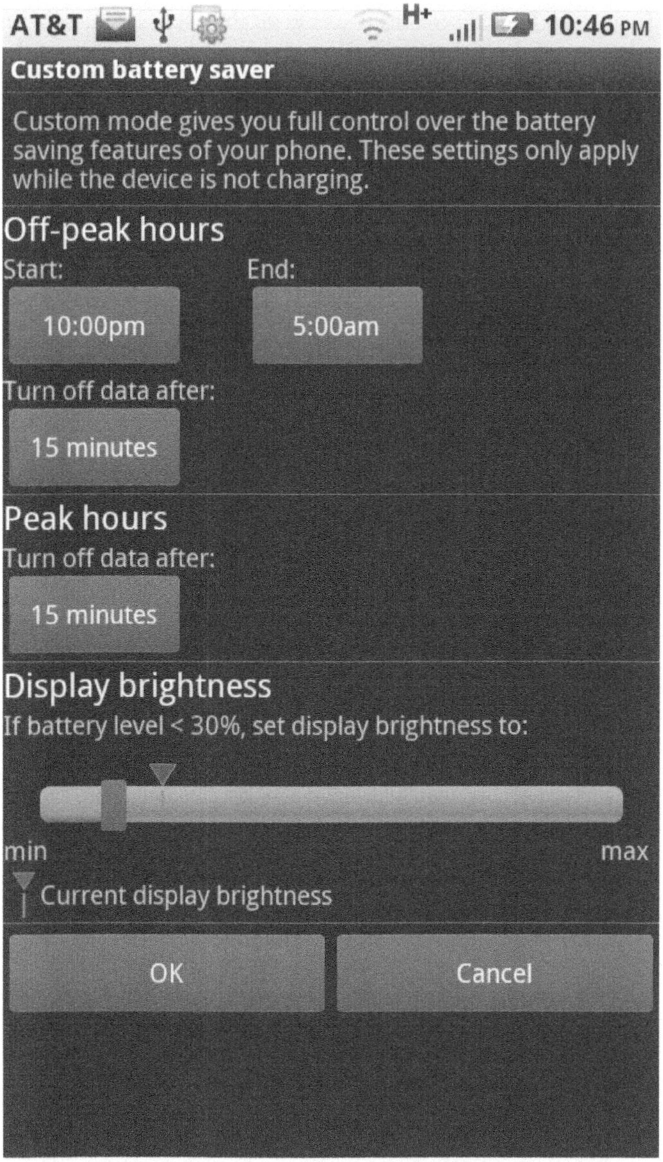

Here you can set the actual Off-peak hours and determine when to turn data off. You may also set the data off option for Peak hours as well. I lowered my setting to 15 minutes. I don't need the phone to constantly attempt to stream data all of the time. I have not seen any interruption or delay in receiving my social network updates, emails, or text messages using the 15 minute setting. The proper setting for you may depend on your own experiences and whether you can be connected to WiFi connections.

The Display brightness setting determines what the brightness should be reduced to when the battery gets low (less than 30%). Since I already lowered my normal brightness (explained in next section), I set the brightness to a bit lower than the current setting.

Select OK once you make your selection choices to return to the Battery Manager screen. Selecting the picture of the battery displays the battery stats:

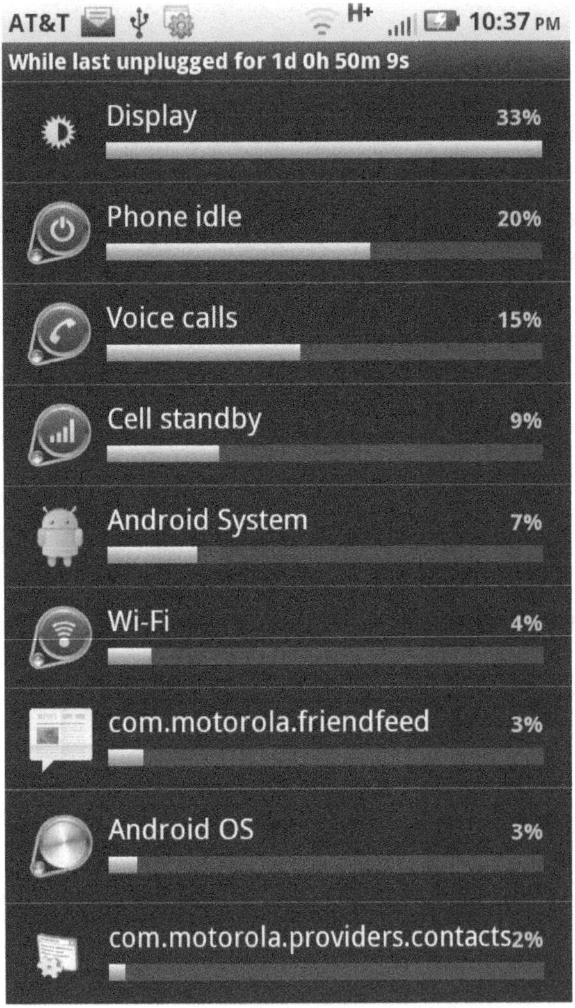

The battery stats provide you insight into what functions or processes are using the most battery power. I noticed when I first starting using my new phone that the Display was using 75% of the battery power - time to change that.

Display Settings

I was surprised to see that most of my battery was being used for display purposes but after thinking about it, most of the work is displaying information. However, I still wanted to tweak that a bit. You can do this right from the Battery Stats screen by selecting the Display entry or from the Display option in the device settings list:

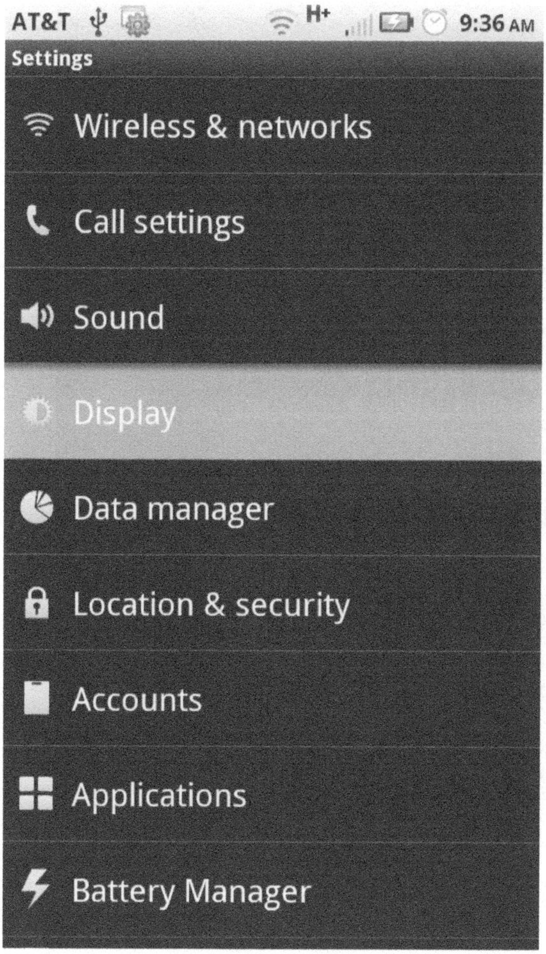

The Display settings screen appears:

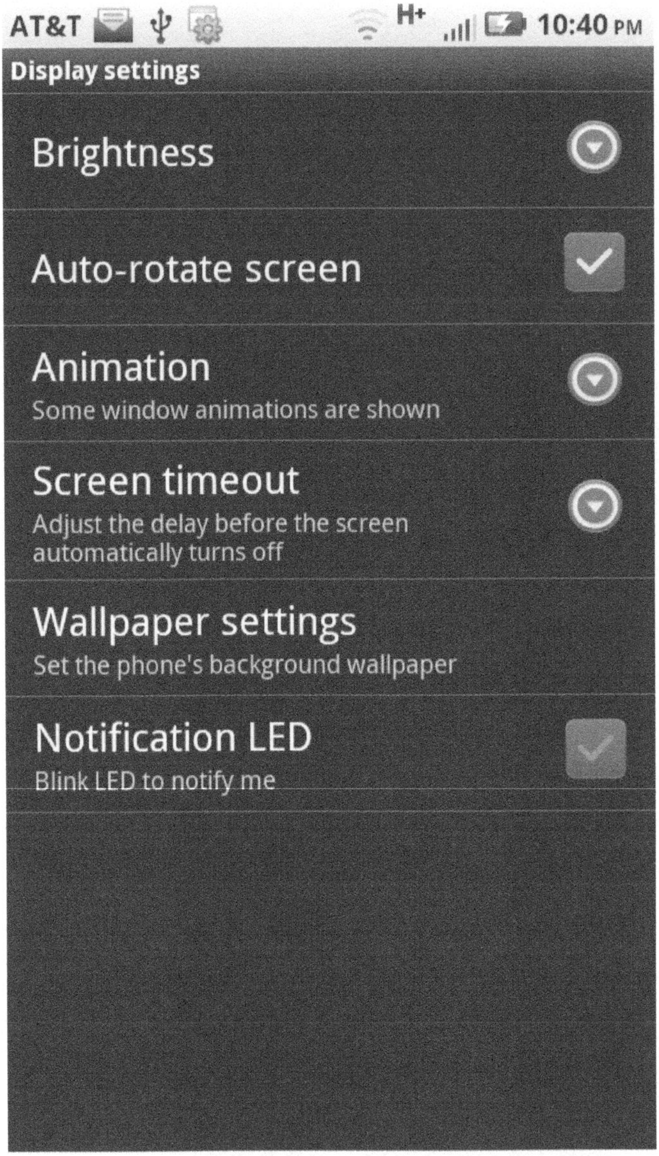

The major setting here is the Brightness. Selecting Brightness displays the Brightness dialog:

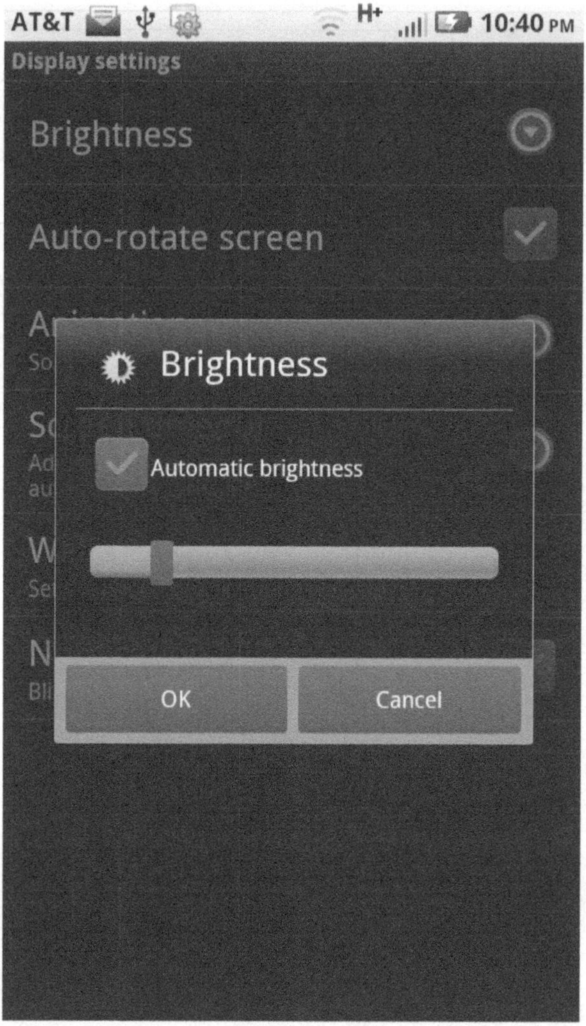

My default setting was Automatic brightness. I liked that but didn't trust it. I therefore unselected the Automatic brightness (checkbox) and adjusted the brightness manually. I was (pleasantly) surprised that I could not tell the difference whether the brightness was all the way up or down to the setting shown in the screenshot. Therefore I set the brightness lower and can't really tell the difference.

The next setting to investigate is the Animation setting. I decided to modify this to only show Some Animations:

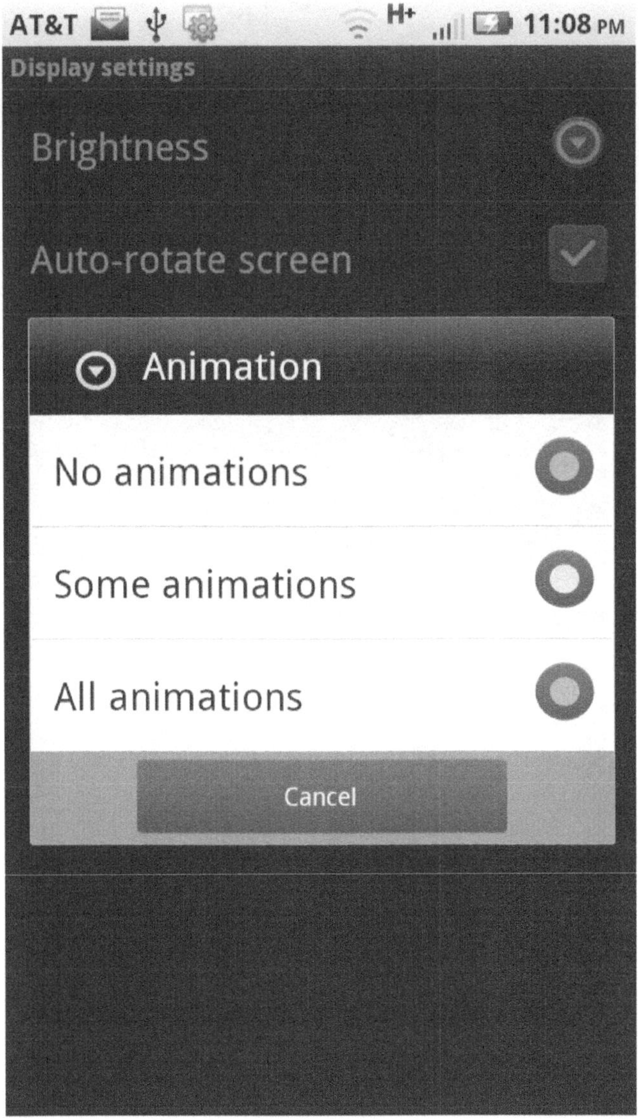

No animations sounds boring but I didn't think I needed All animations if my battery life is at stake. Therefore, I thought Some animations is a nice happy medium.

Next, is probably the best (or worst) setting that helps save battery life - Screen timeout:

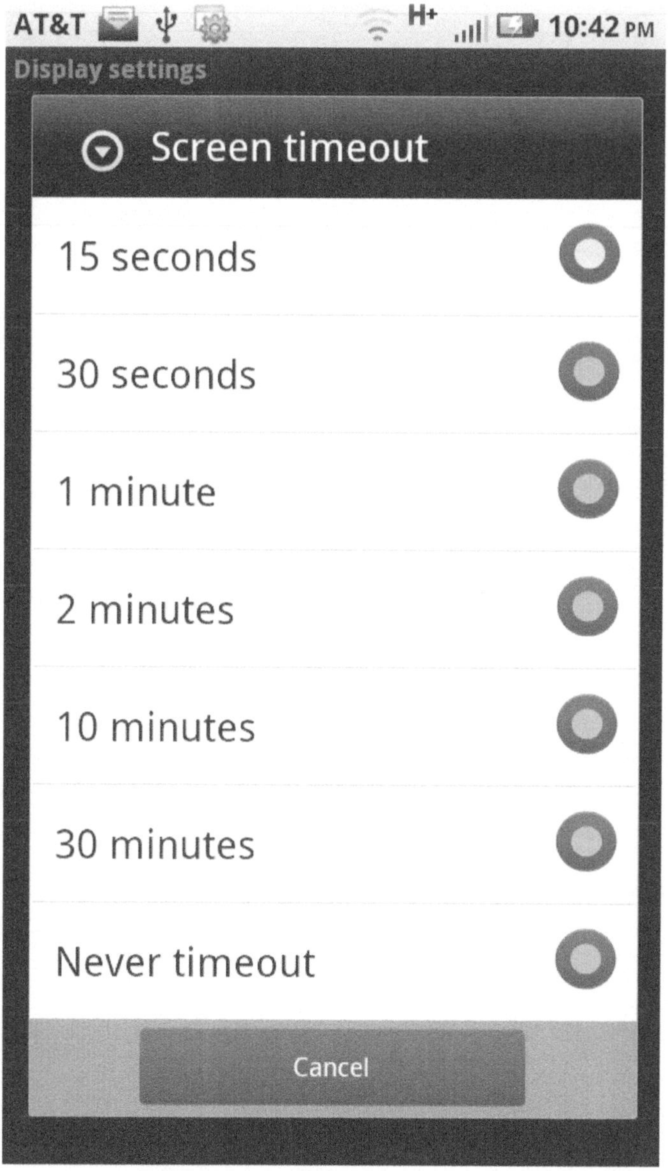

I was daring and changed the timeout to 15 seconds. Sometimes this is perfect, other times it is annoying. I may need to change this to 30 seconds. You should play with this setting and find what is right for you. For example, it takes me less than 15 seconds to initiate a voice call, and then once connected (usually through my Bluetooth), I don't really need the screen to be on (well at least not when I am talking to my wife - I'd better be paying attention to the call and not surfing the web or checking email).

Finally, I turned off the Notification LED setting. I really don't pay attention to the LED and if it was flashing anyway, I wouldn't really see it until I was checking my phone anyway - therefore I didn't see any reason to have the LED wasting any power (albeit probably a small amount).

Sound Settings

Your Sound settings also play a part in battery usage. From the device settings menu, select Sound:

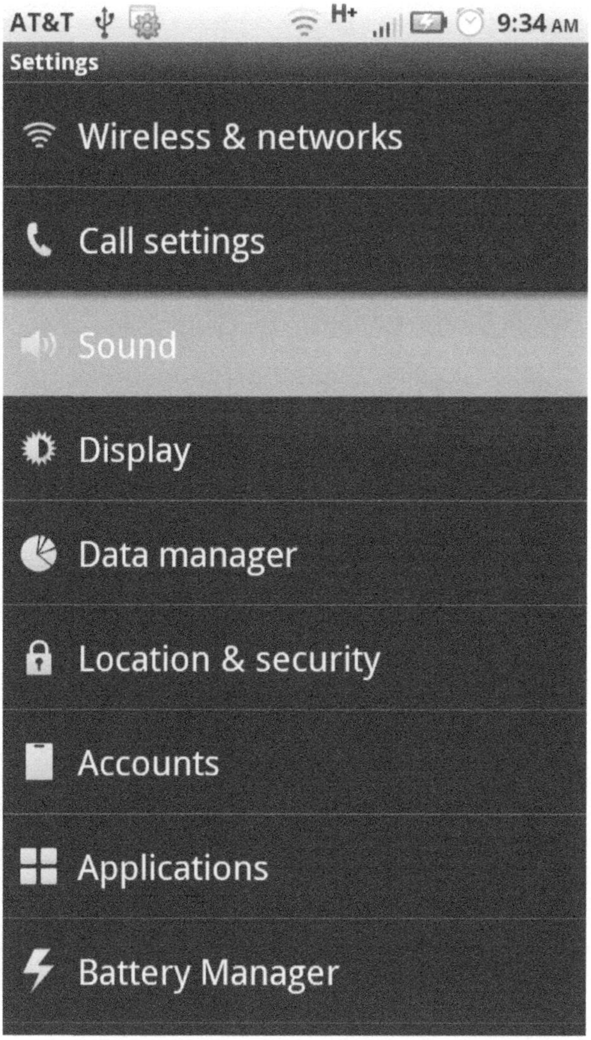

The Sound settings screen displays as shown:

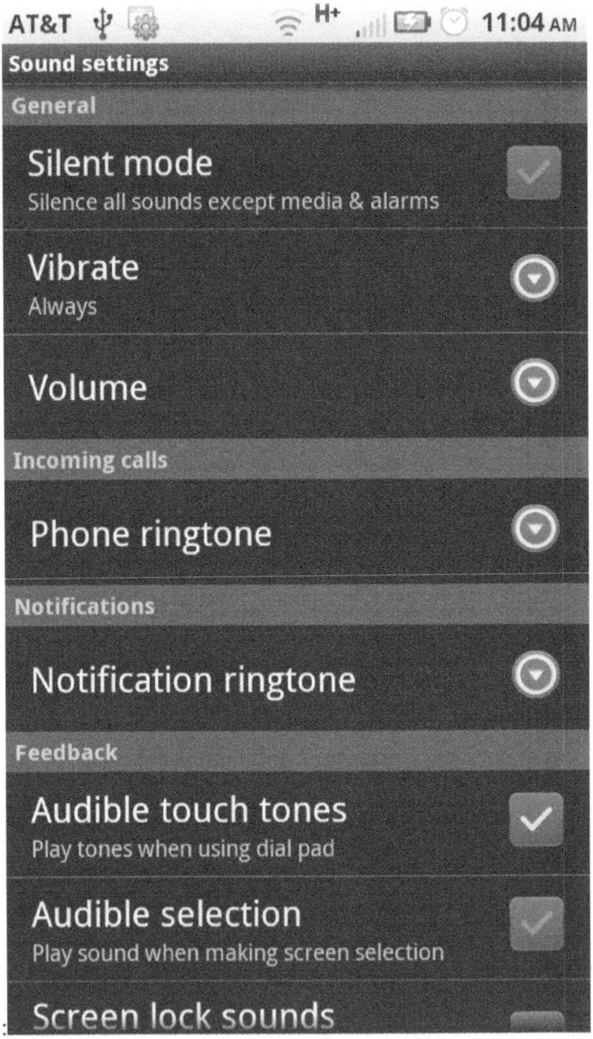

Selecting Vibrate opens the Vibrate setting dialog. Having your phone vibrate requires more power than if it doesn't vibrate. Sometimes I don't hear my phone so I need to have it vibrate at all times. However, you may decide to select Never or Only in Silent mode to reduce the need for vibration power.

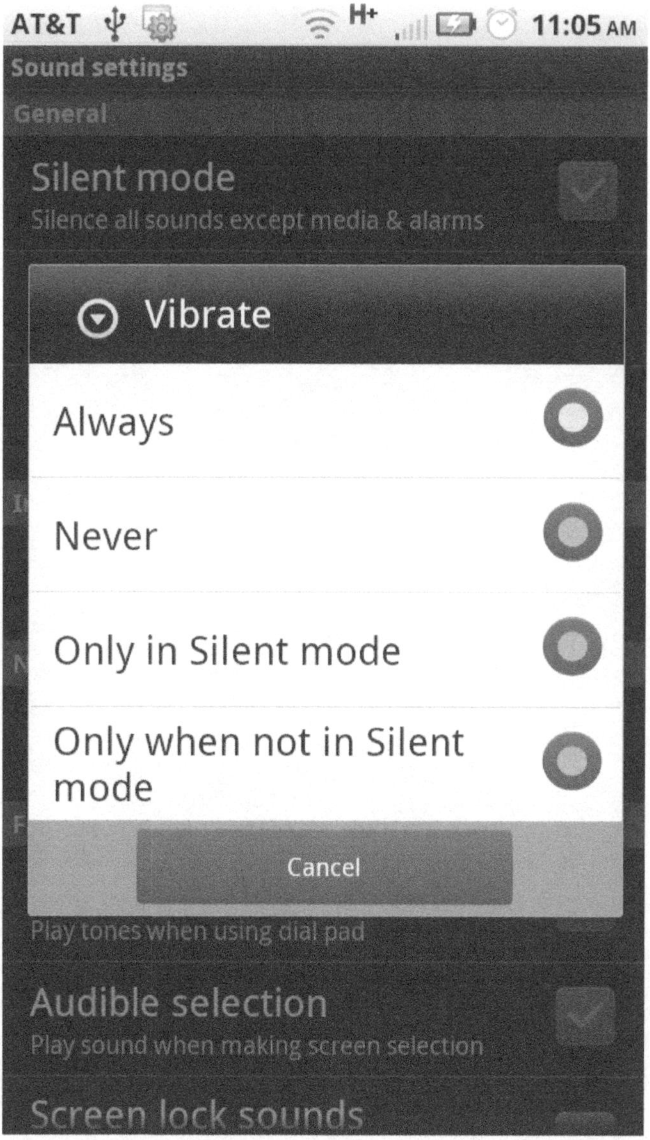

The next sound option is Volume. Selecting Volume opens the Volume dialog window. The louder the settings here, the more power is being used. I don't need to cause attention by having the loudest settings. Toning the settings down a bit will reduce an embarassment and help conserve battery power.

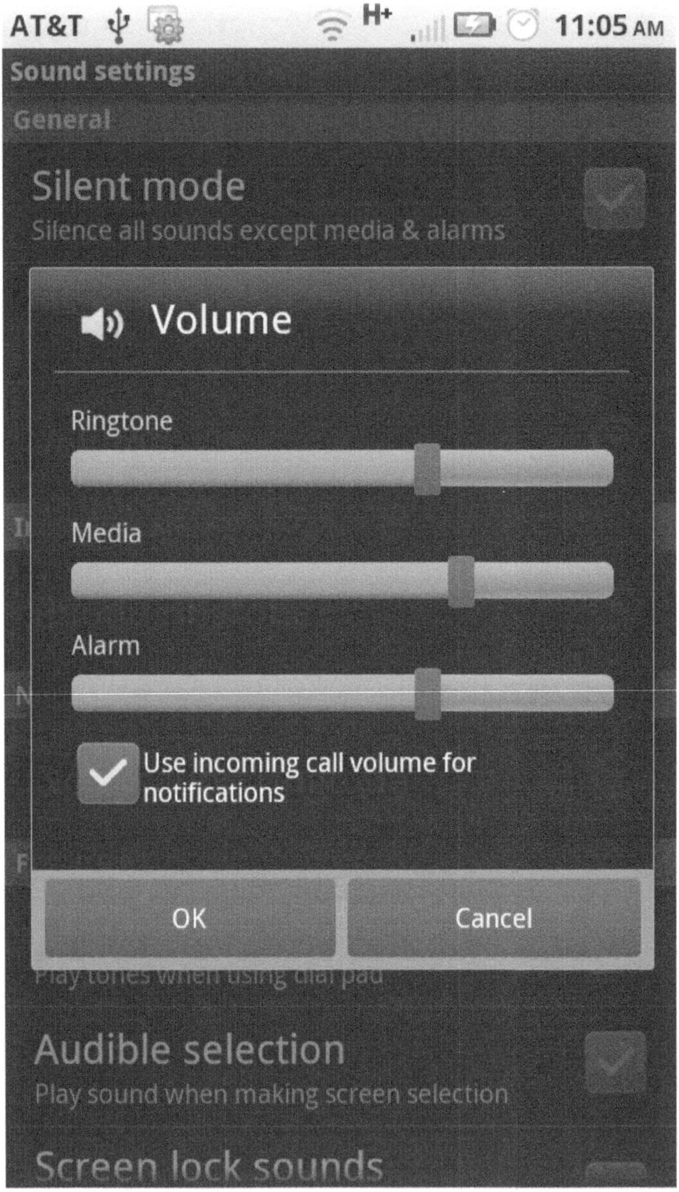

The next settings from the Sound settings menu are the ringtones for phone calls and notifications. While I used to think having my favorite hip-hop song as a ringtone was cool, it doesn't work so well in the business world. I keep my ringtone settings set to built-in ringtones that are simple sounds and not long song-like melodies or downloaded ringtones. By having quaint ringtone settings, you phone uses less power to play them and thus helps conserve battery power.

WiFi Settings

The WiFi settings have impact on battery life but deciding your strategy depends on available WiFi networks throughout your day and your data transfer preferences (which are explained more under Messaging and Data Manager).

You may modify your WiFi settings by selecting Wireless & networks from the device settings menu:

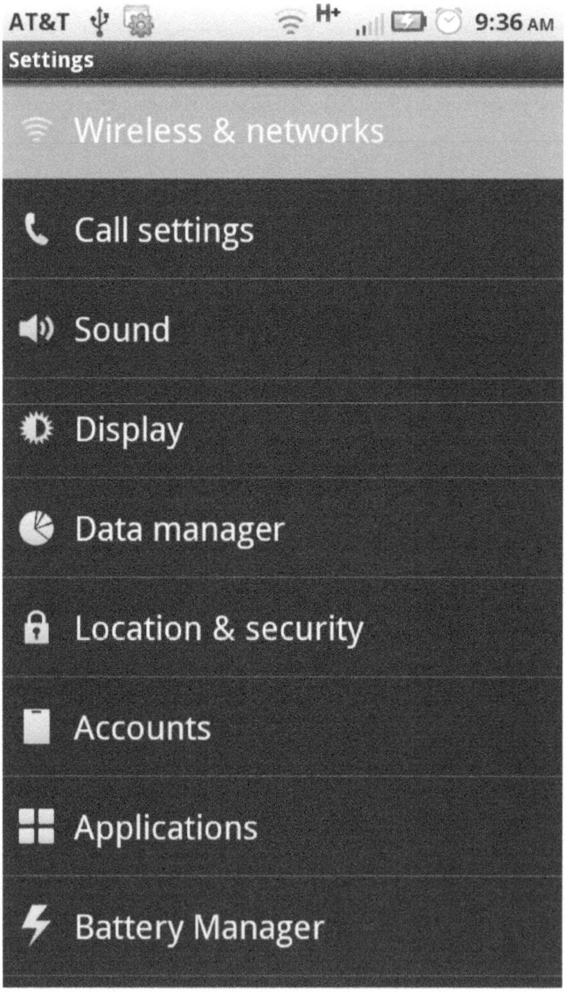

The Wireless & networks setting screen appears:

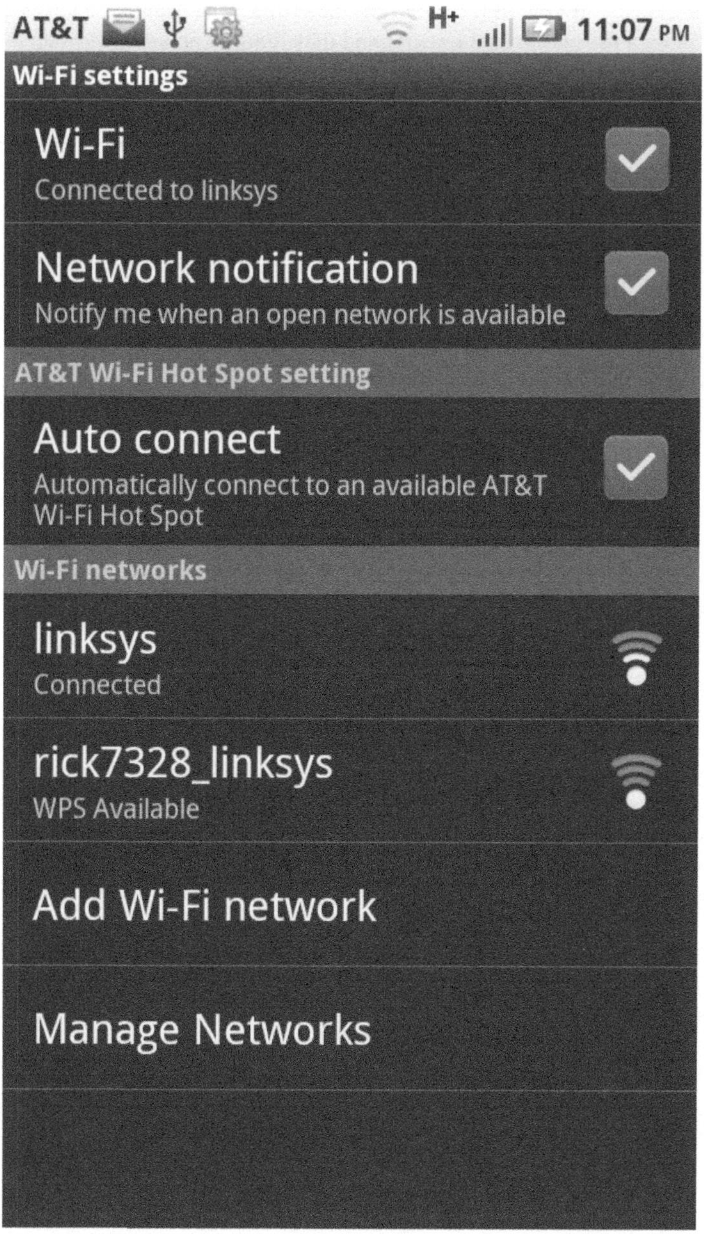

There are recommendations to turn WiFi off when you know you won't be connecting to a WiFi connection. This also stops the device from searching and notifying of available WiFi signals. However, in my case, I am almost always near a WiFi connection whether it be my home wireless network, my work network, my internet provider hotspot, my phone provider hotspot, or some random guest connection.

Therefore I keep WiFi on all the time. The data transfer is usually faster via my WiFi connection than the phone data transfer and thus reduces power usage and data plan usage to retrieve updates and messaging.

The network searching and notification can drain power but if you are in the same or similiar locations each day, everytime you get a notification of an available network or hotspot, connect to that signal (assuming you have access) and the device will remember the network. Your WiFi connection will therefore connect to any available remembered networks without the need to notify you.

BlueTooth Connection

I like using a BlueTooth with my phone so I am hands free and not holding such a powerful device up to my head (not that there is anything proven from that but the discussions of radiation can get you nervous). I also need to use it while driving.

Since I only use the BlueTooth for voice calls which I do not make all the time, I don't need the BlueTooth connection running and operating on the phone all the time. Therefore I only turn the BlueTooth connection on when I am on a phone call and then turn it off when I am done.

Some phones have quick connection widgets where you may easily turn the various connections on and off:

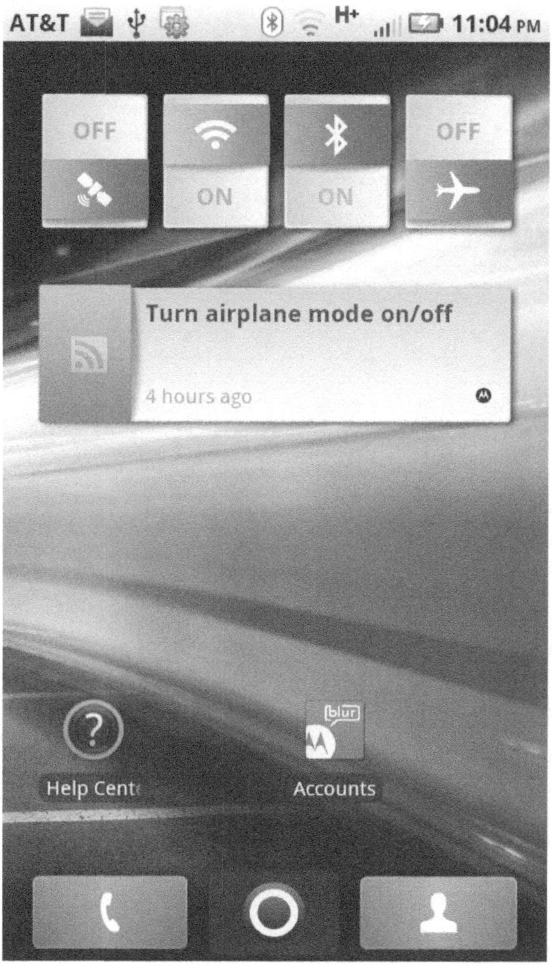

Data Manager

You may tweak your data delivery options by using the Data Manager from the device settings menu:

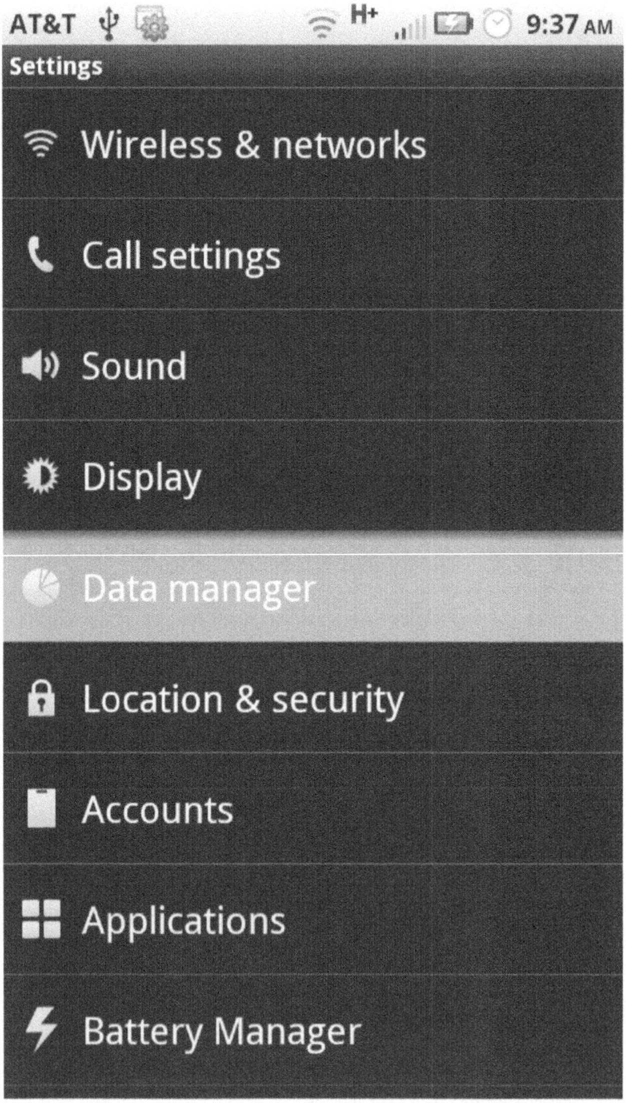

Here you get the Data Manager screen:

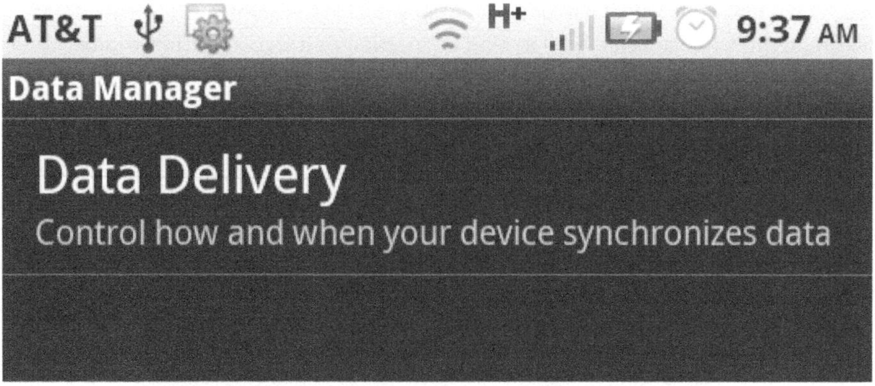

Select Data Delivery to display the Data Delivery options:

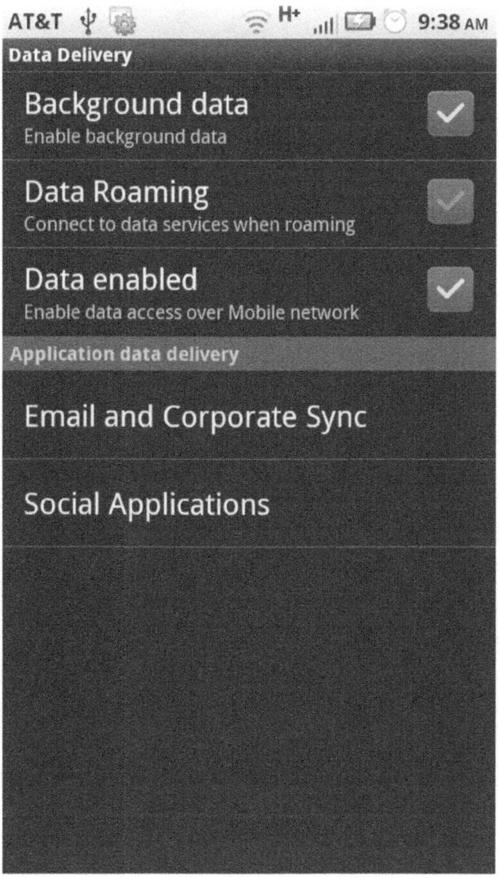

While Background data will use more power, I like to receive any messages, emails, etc. when I can so I leave that alone. You may decide to turn this off. I leave Data Roaming off as I can usually hit a WiFi connection or data connection without problems. If you have dificulty connecting you may need to activate Data Roaming. Of course I leave Data enabled on as what is the sense of having a smart phone device if you can't receive data.

The next section of the Data Delivery options screen are the Application data delivery items. Select Email and Corporate Sync to tweak your email options:

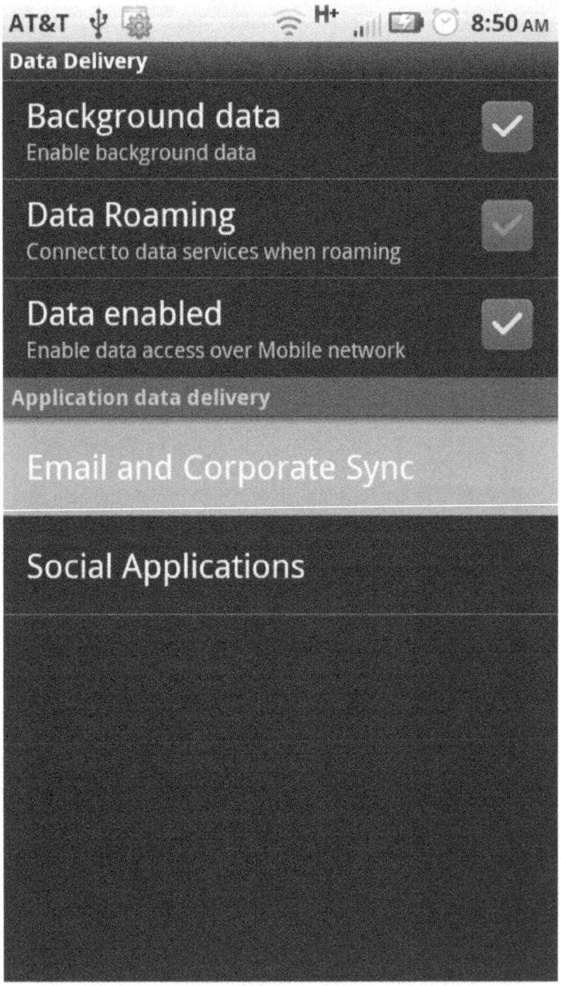

The Email Settings screen appears:

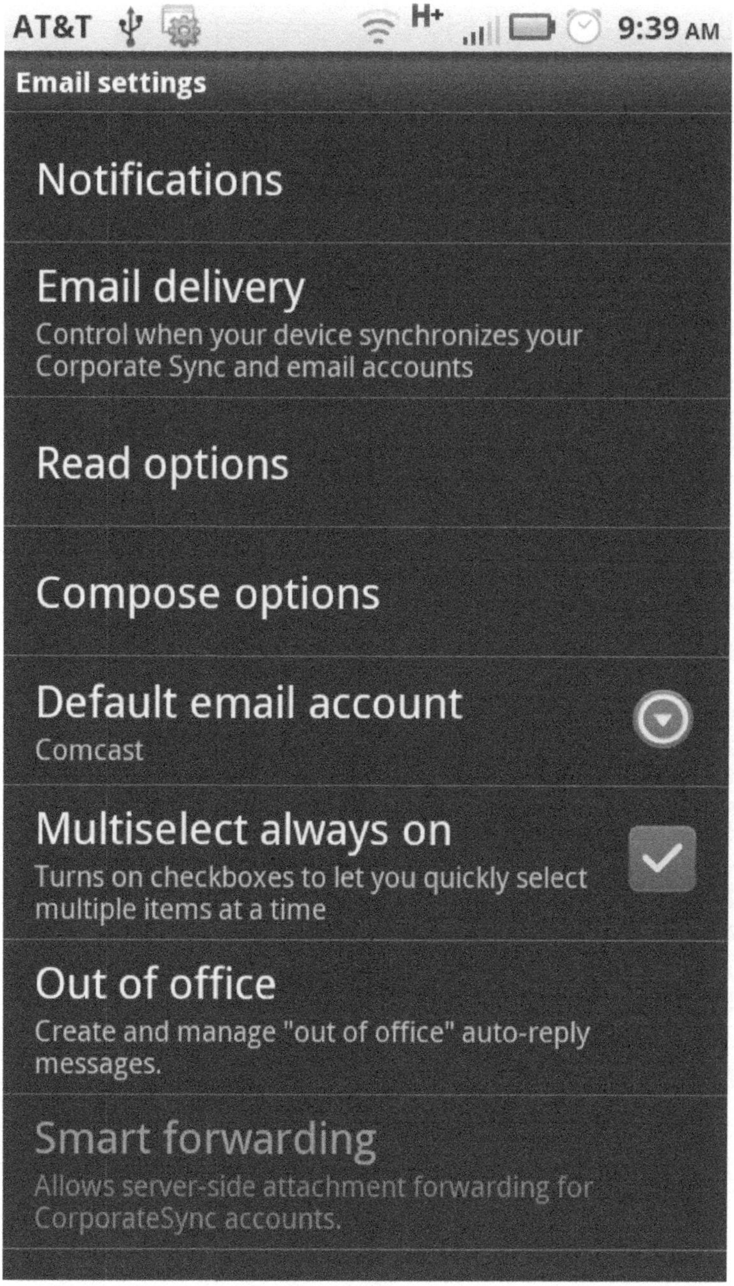

The key selections here are the Notifications and Email delivery. Select Notifications to go to the Email notifications screen.

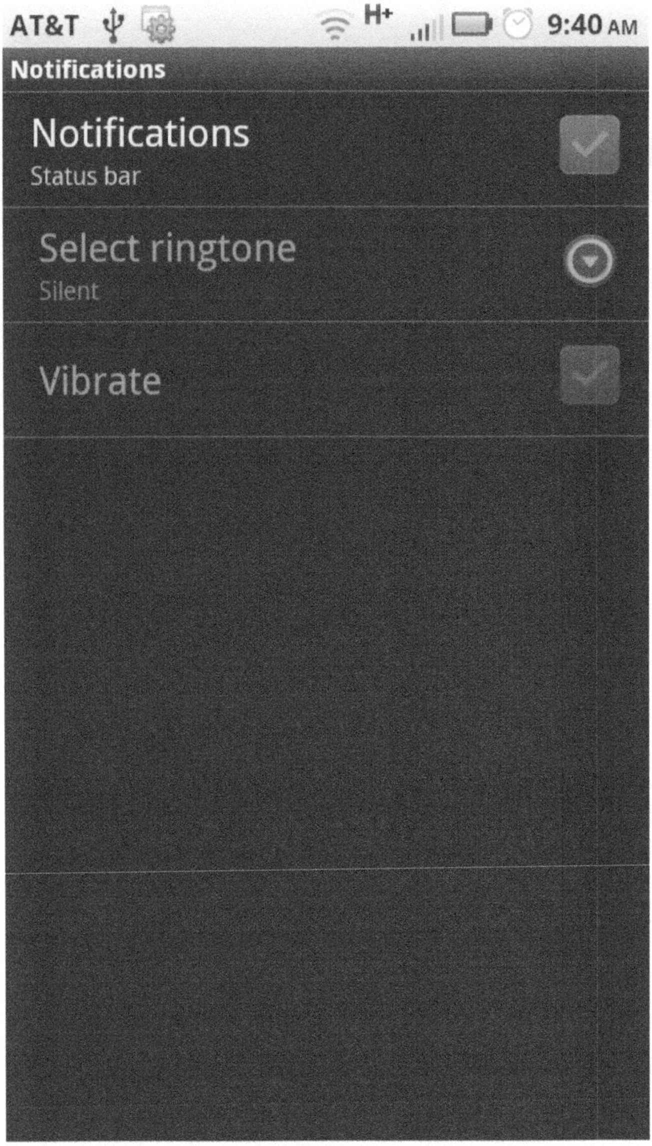

Most of the time, my phone is on my side within its holder. I therefore don't need a notification on the status bar since I won't see it anyway. Therefore I turned that off. You may select a ringtone or vibrate option here but I leave them off. I usually check my email during breaks and lunch so I don't need to be notified each time I get a new email. Some people have these set on but they are just showing off that they are getting emails. It can be annoying during meetings to see or hear people get emails and not pay attention to the discussions. Turning these off will indeed save power.

The next setting is the Email Delivery option from the Email Settings screen. Select Email Delivery to go to the Email Delivery screen.

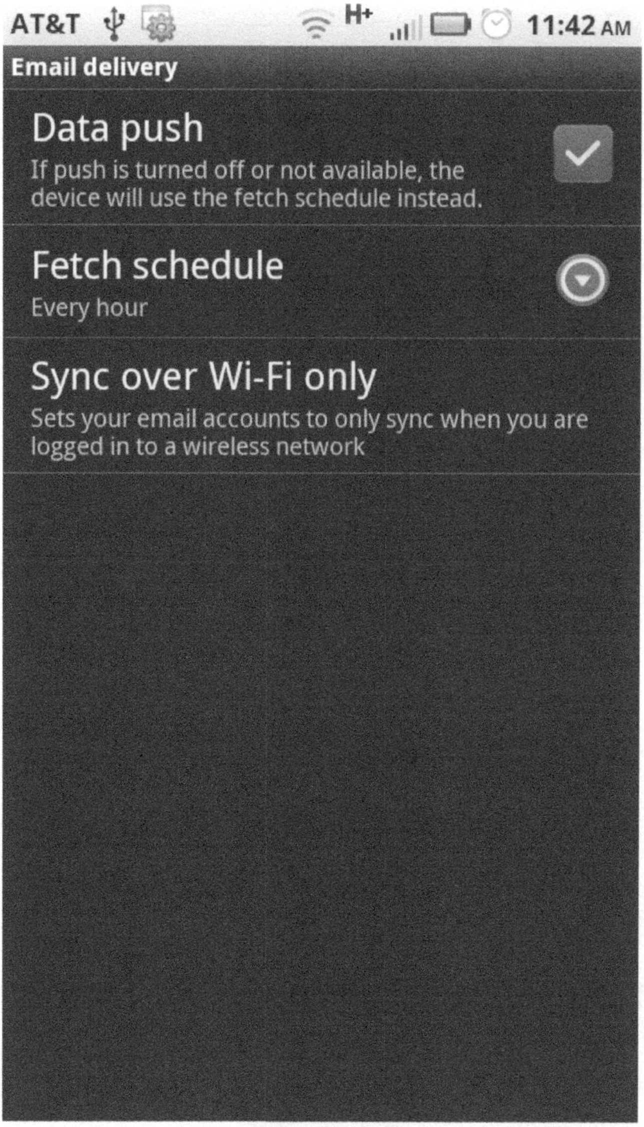

The key setting here is the Sync over WiFi only. You may want to turn this on if you are constantly connected to WiFi as that is quicker than using the data transfer from your device. It will use less power and less of your data plan.

Back on the Data Delivery menu, select Social Applications:

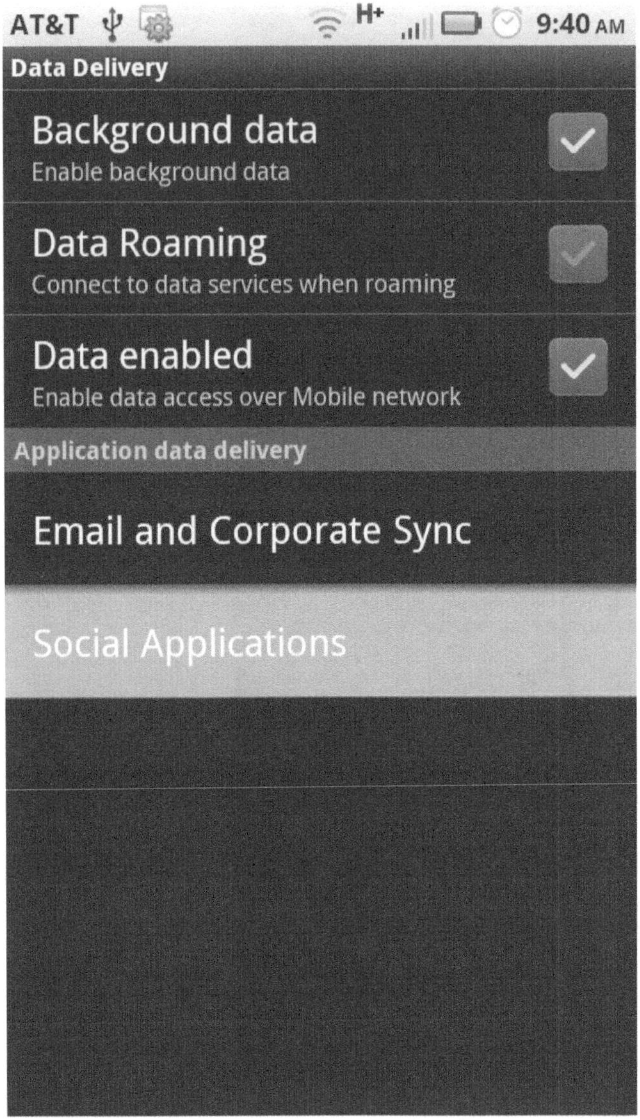

This brings up the Social Applications screen.

The option here is the Sync over WiFi only. Again, you may want to turn this on if you are constantly connected to WiFi as that is quicker than using the data transfer from your device. It will use less power and less of your data plan.

Messaging Settings

In addition to the Data Delivery options, the messaging settings on your phone tweak similiar settings (in fact for email they are exactly the same). Find the Messaging app within your applications menu screen:

Running the Messaging app displays the Messaging screen:

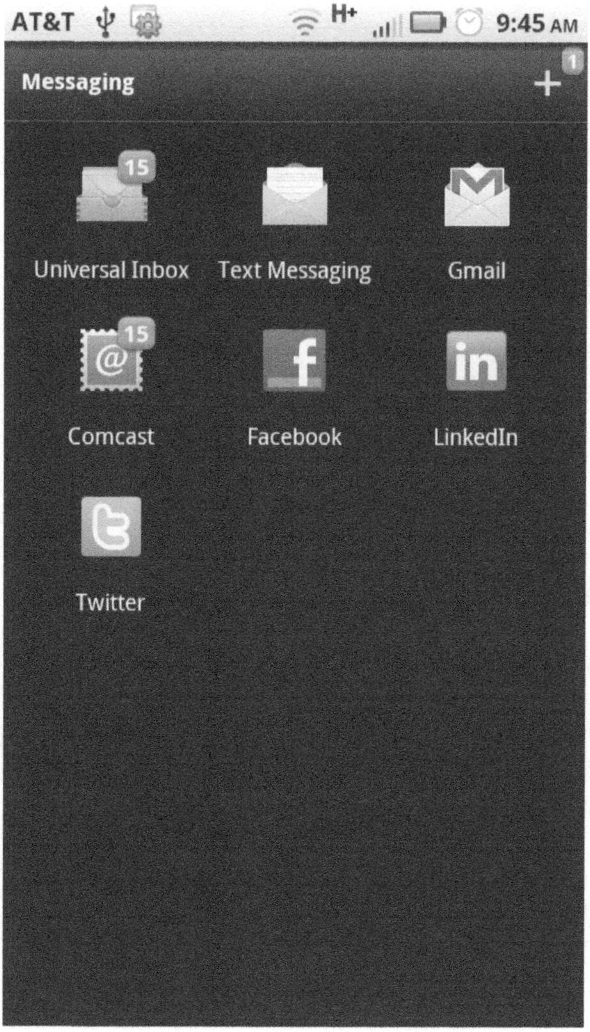

Use your device's settings button to select the Messaging Settings.

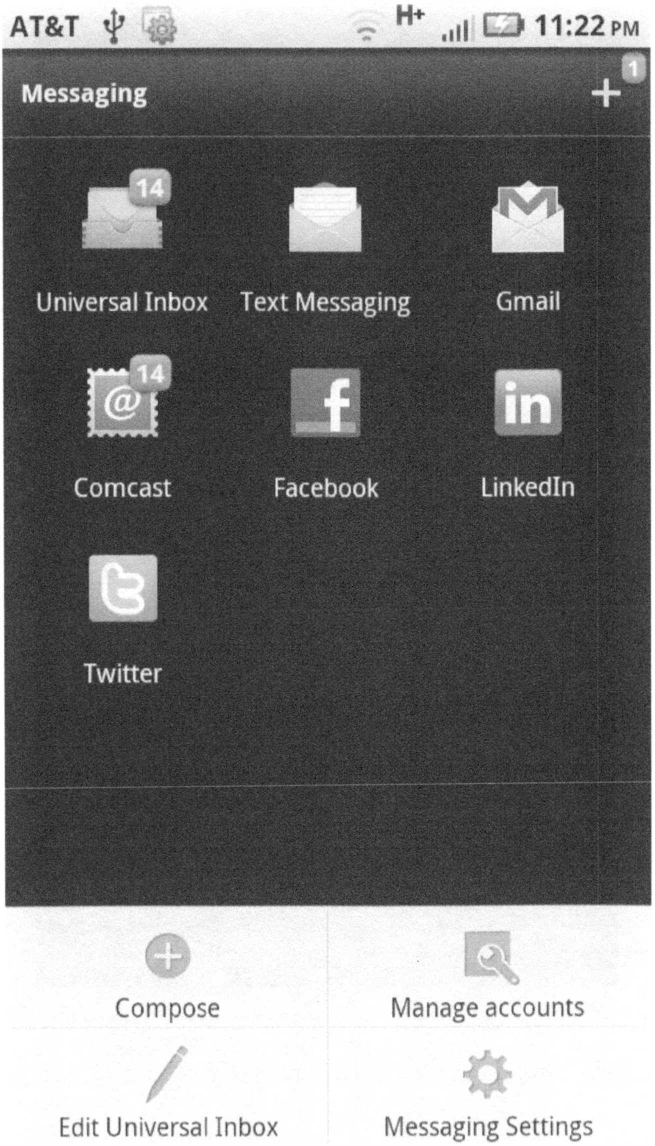

This brings up the Messaging Settings menu:

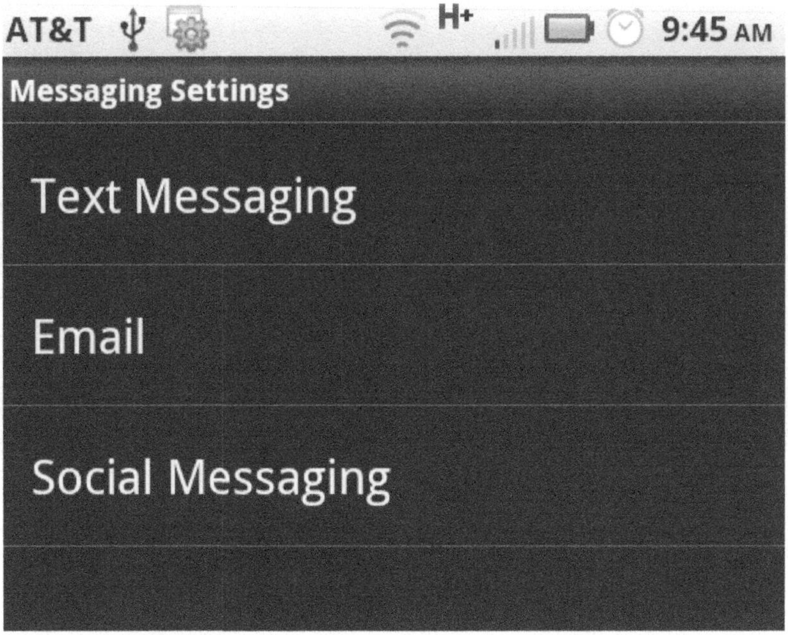

Select Text Messaging to modify the text messaging settings.

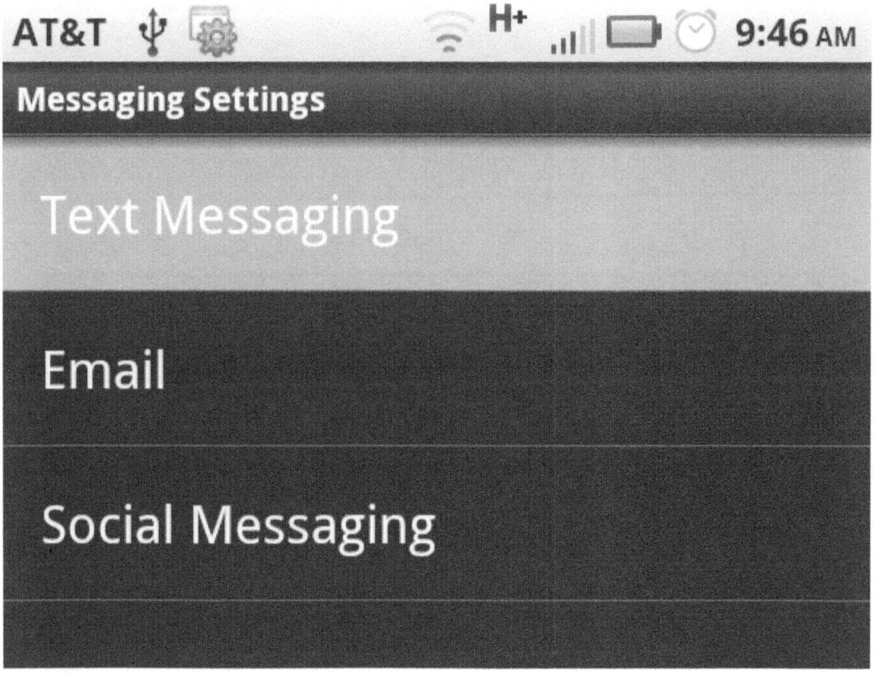

Within the Text Messaging settings you may change the Notifications and the Ringtones. I leave these settings alone as I want to know when I get a text message as it is usually family related.

Back on the Messaging Settings menu, select the Email option:

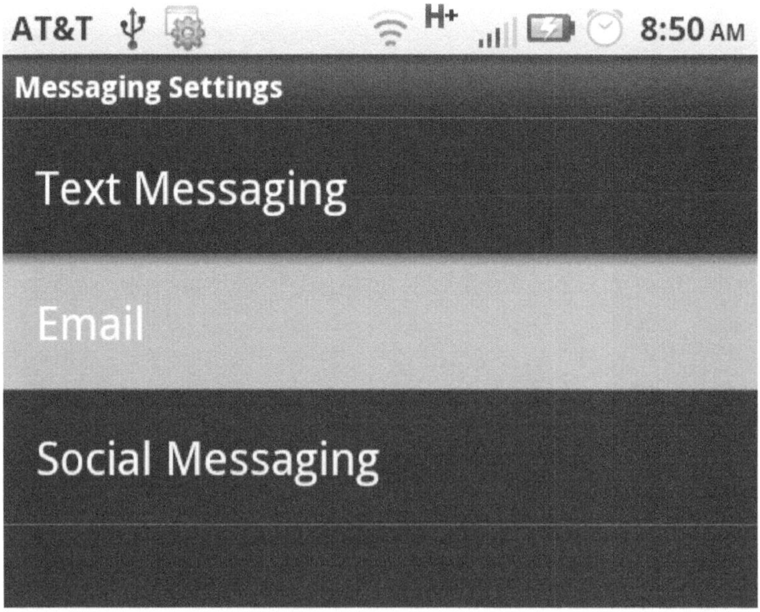

The Email Settings screen appears. This provides the same settings and menus as the Data Delivery options within the Data Manager. The options are repeated here so you don't need to refer back to the Data Manager section.

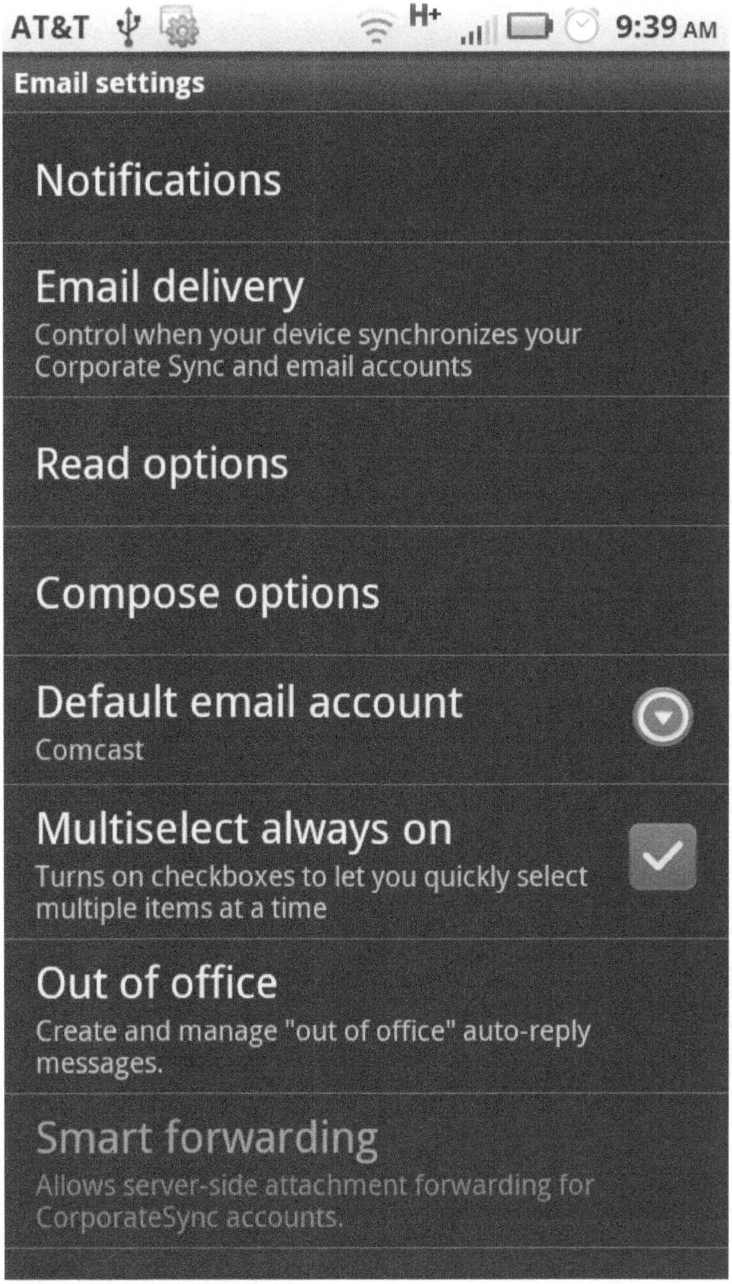

The key selections here are the Notifications and Email delivery. Select Notifications to go to the Email notifications screen.

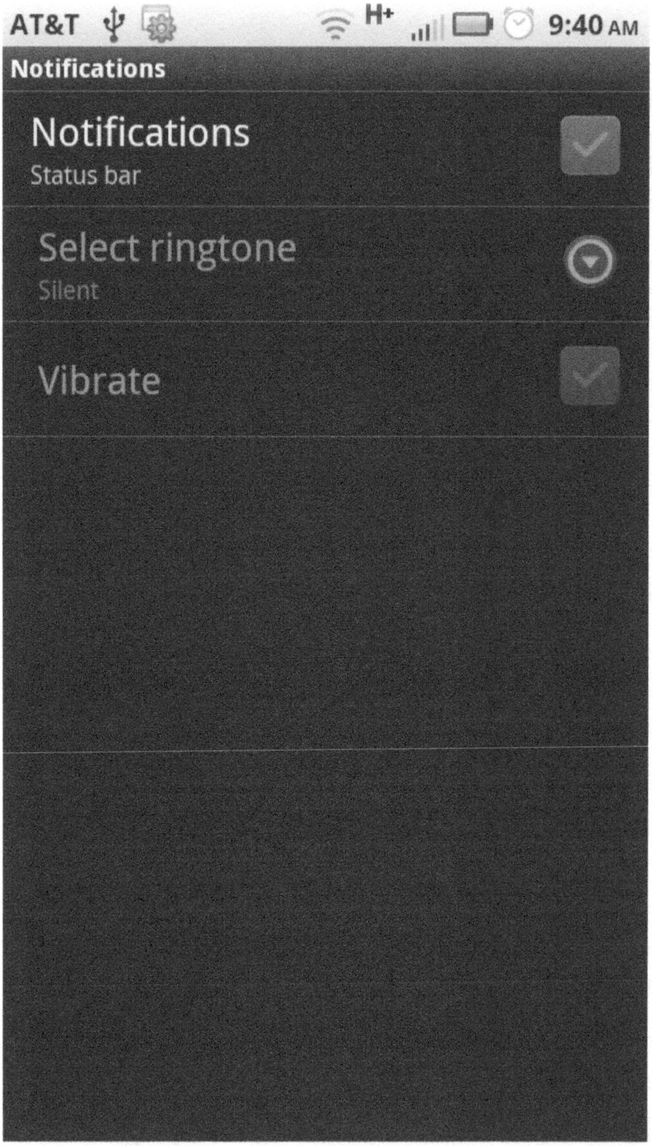

Again, most of the time, my phone is on my side within its holder. I therefore don't need a notification on the status bar since I won't see it anyway. Therefore I turned that off. You may select a ringtone or vibrate option here but I leave them off. I usually check my email during breaks and lunch so I don't need to be notified each time I get a new email. Some people have these set on but they are just showing off that they are getting emails. It can be annoying during meetings to see or hear people get emails and not pay attention to the discussions. Turning these off will indeed save power.

The next setting is the Email Delivery option from the Email Settings screen. Select Email Delivery to go to the Email Delivery screen.

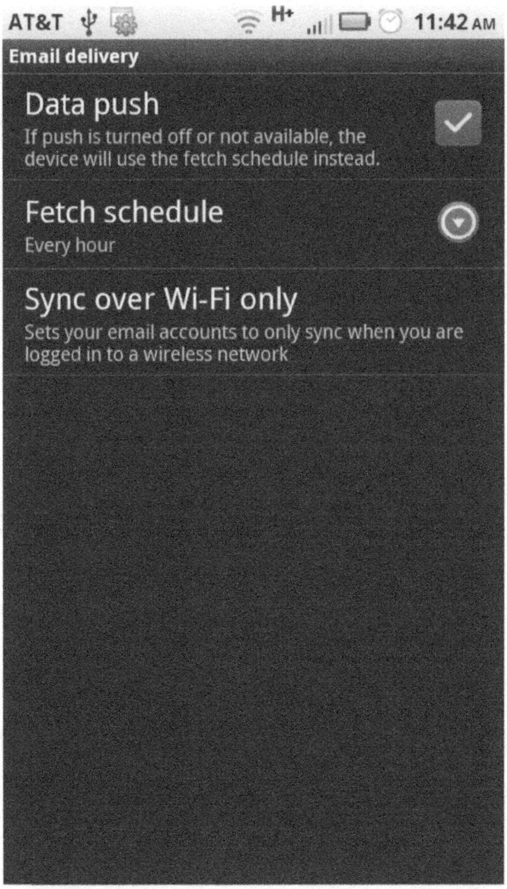

The key setting here is the Sync over WiFi only. You may want to turn this on if you are constantly connected to WiFi as that is quicker than using the data transfer from your device. It will use less power and less of your data plan.

Back on the Messaging Settings menu, select Social Messaging:

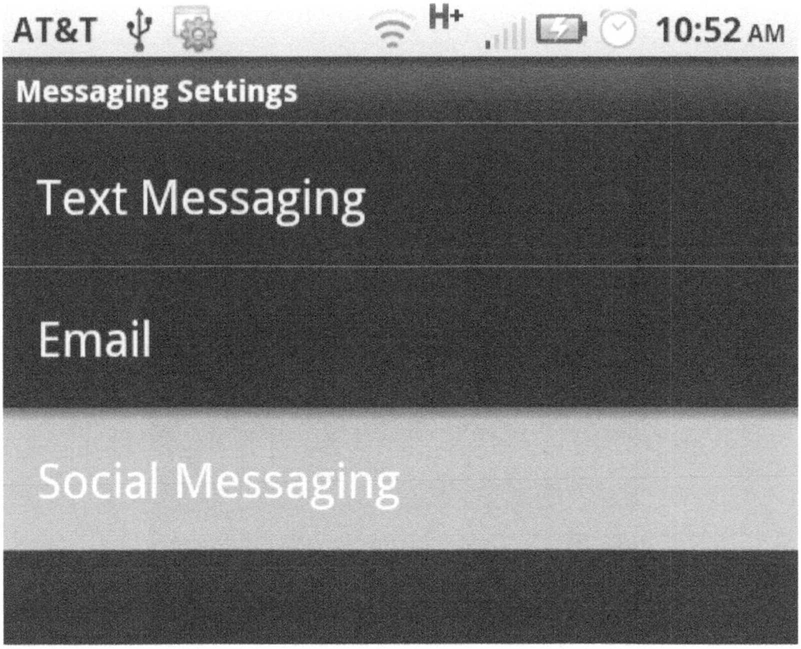

The Social messaging settings screen appears:

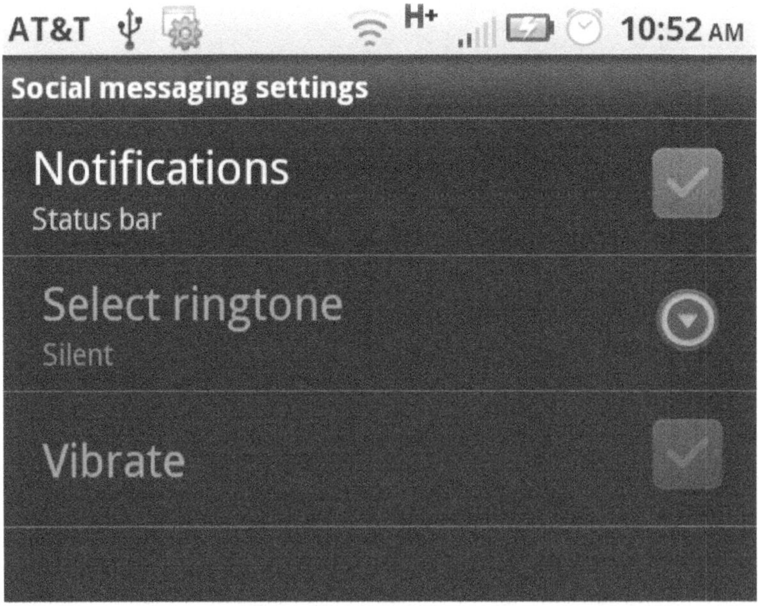

Since I review my social messages in batches, I don't need to be notified each time an update comes through. Therefore I turned off all notifications, ringtones, and vibrate options. This definitely saves power for me. I get so many status updates which most I do not care about.

Apps on Phone Top

It is a great idea to place frequently used apps onto your phone top. That way, you can easily get to the apps you use most without having to search for them within your applications menu. This saves a bit of time and energy which eventually adds up.

Task Manager

After running your phone apps including the web browser, the applications may still be running in the background. Just because you end your browser session or Twitter app, for example, doesn't mean it is closed.

Depending on the application, even though it may be idle, power may be used for logging, monitoring, and internal notifications. Therefore it is key to use the Task Manager app to tweak your applications.

Some phones may have a Task Manager but if not, there are free task manager apps available from the Android Market. Select the Task Manager app from your applications menu:

The Task Manager app is launched:

AT&T
H+
10:57 PM
Task Manager
Applications
Auto-end list
Browser
CPU: 0% RAM: 16 MB
Twitter
CPU: 0% RAM: 11 MB
LinkedIn
CPU: 0% RAM: 8 MB
Vlingo
CPU: 0% RAM: 4 MB
AT&T Navigator
GPS
CPU: 0% RAM: 3 MB
Gmail
CPU: 0% RAM: 3 MB
MediaShare
CPU: 0% RAM: 3 MB
AT&T Code Scanner
(not running)

Any app listed with a CPU entry is running. Although you can force these to end, you may want to have your apps set to Auto-End. Select the apps that you would like to set to Auto-End and then select the Add to Auto-end button at the bottom of the screen:

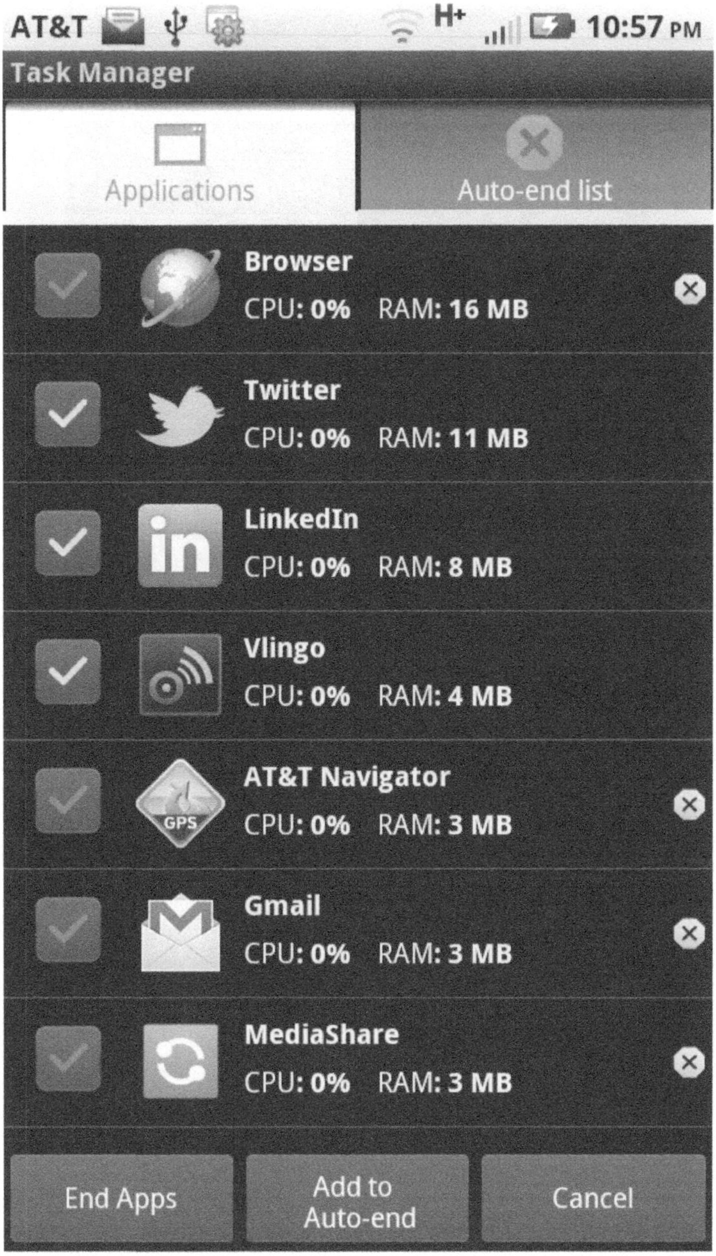

You may also force apps to end by selecting them and then selecting the End Apps button:

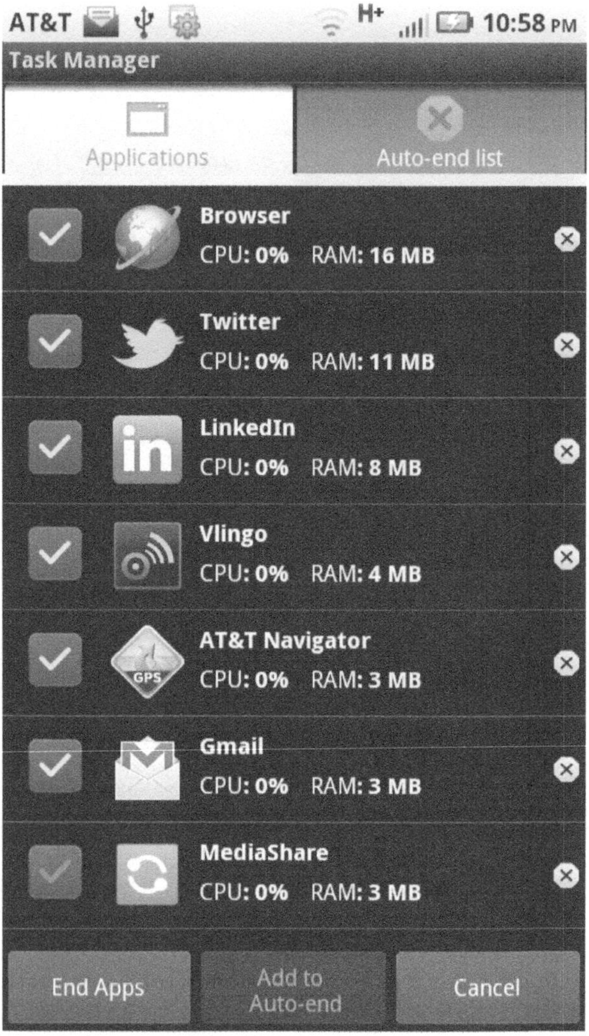

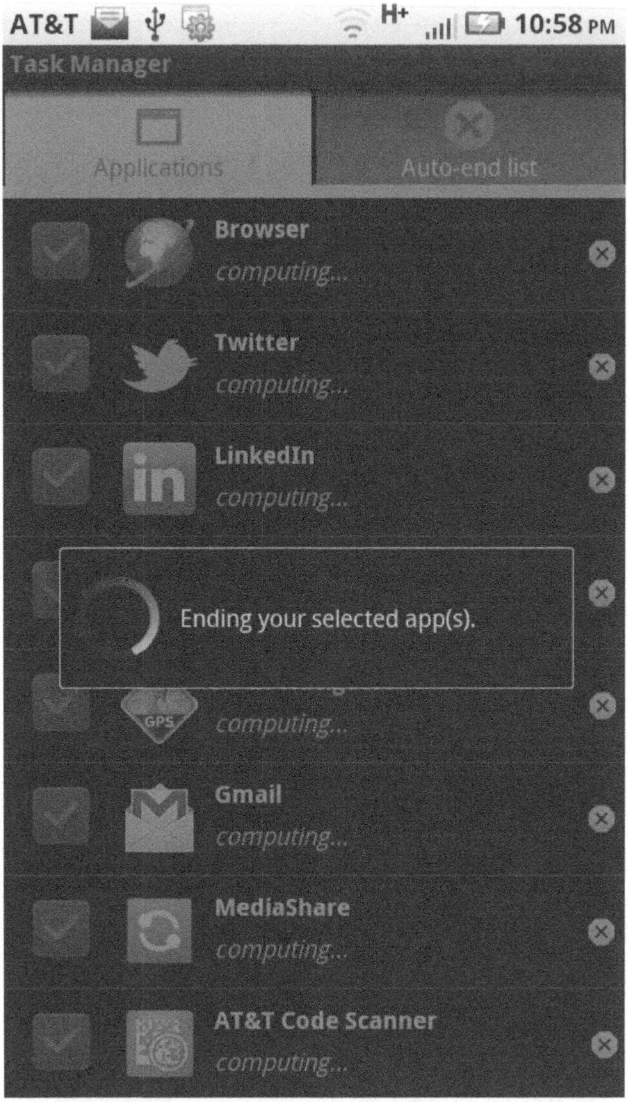

This makes sure that Apps are still not running in the background when you are not using them - sucking up battery life for no reason.

Application Settings

Select the Applications option from your device settings menu:

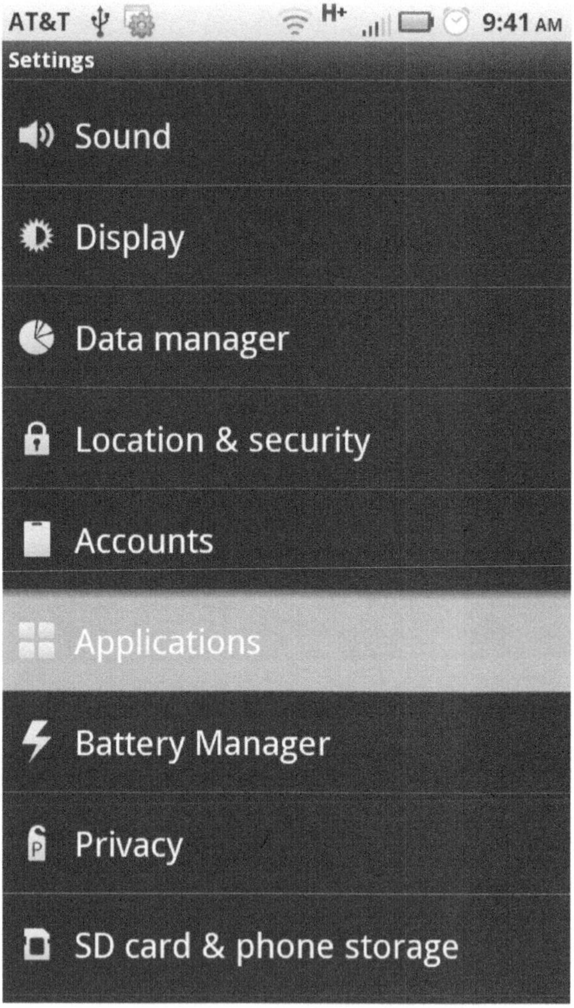

The Application Settings screen appears:

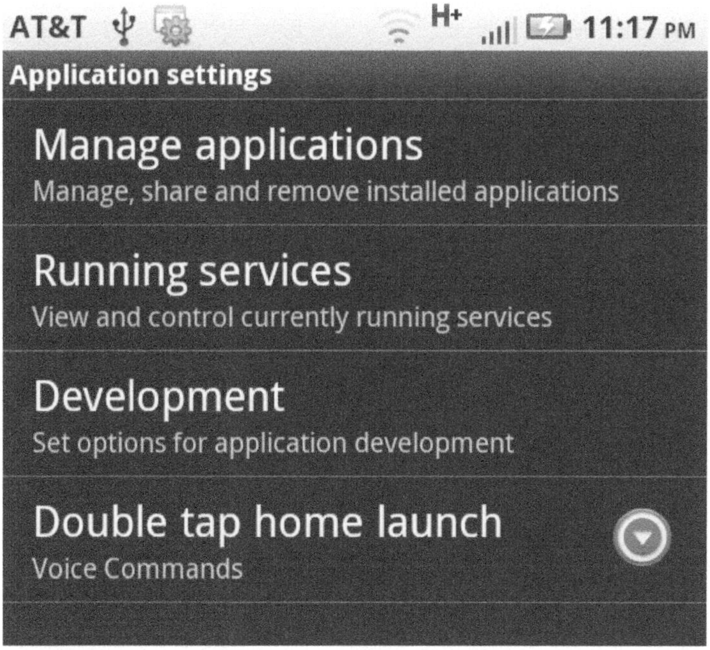

The Running Services is similar to the Task Manager but actually shows the services running from your applications:

It is probably safer to end your applications than the underlying services, however, if you feel comfortable doing so, you may end the services here.

The Development options from the Application Settings menu should all be turned off unless you are performing Android development. These options will use more battery power if activated.

Rebooting

Rebooting your phone insures that it is refreshed and all apps and services that are running are shut down - just like a regular computer or laptop. It is a best practice to reboot your phone at least once per day. I usually shut it down before I place my phone on the charger. I let the phone charge while it is off. It is good to give it a break. To power off your phone hold down the power button and then select Power Off from the menu:

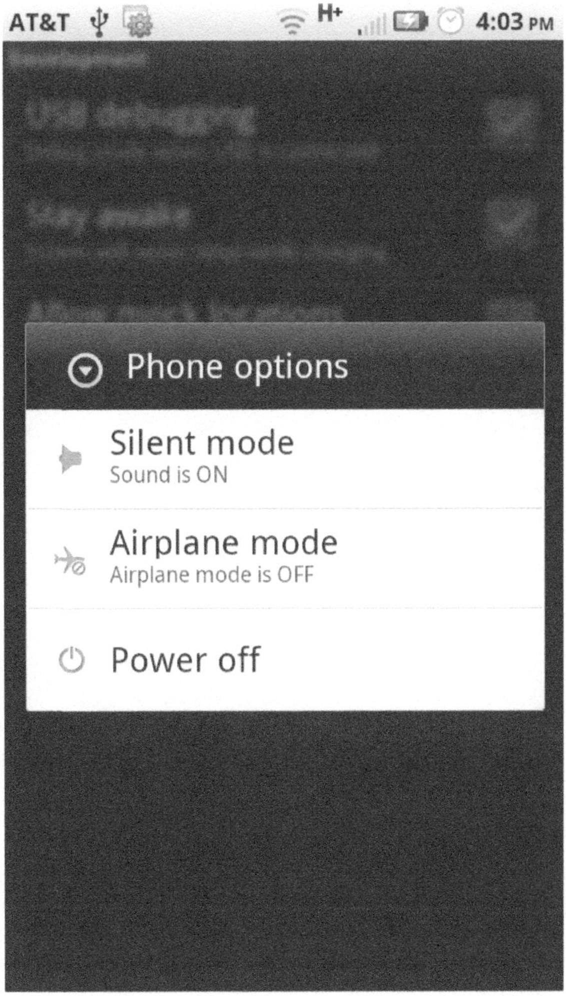

Conclusion

There are many options and settings on your Android powered smart phone that affect battery life and performance. There is always a balance (or struggle between the two). However, modifying your settings ever so slightly may improve your battery life. You need to decide how you use your phone and what makes the most sense for based on your usage, connections, preferences, etc. No matter what, I am sure you found a few things that you could tweak to extend your phone's power throughout your day.

About the Author

Steve Mann was born and raised in Philadelphia, Pennsylvania, where he still resides. He is Enterprise Applications Engineer for Morgan Lewis and has more than 16 years of professional experience. Steve was previously a Principal Architect for RDA Corporation where he worked for over 13 years. He has authored and co-authored several books related to collaboration technology.

For the past 7 years, he has primarily focused on collaboration and business intelligence solutions using Microsoft technologies. Steve managed the internal BI Practice Group at RDA for several years. He was also heavily involved within RDA's Collaboration/Search Practice Group.

Steve's blog site can be found at: www.SteveTheManMann.com

Made in United States
Cleveland, OH
04 November 2026

47644153R00036